SMART WORDS READER

You've Got Guts

The Digestive System

Christine A. Caputo

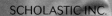

SCHOLASTIC INC.

What are SMART WORDS?

Smart Words are frequently used words that are critical to understanding concepts taught in the classroom. The more Smart Words a child knows, the more easily he or she will grasp important curriculum concepts. Smart Words Readers introduce these key words in a fun and motivational format while developing important literacy skills. Each new word is highlighted, defined in context, and reviewed. Engaging activities at the end of each chapter allow readers to practice the words they have learned.

ISBN 978-0-545-33446-4

Packaged by Q2AMedia

Copyright © 2011 by Scholastic Inc.

Picture Credit: t= top, b= bottom, l= left, r= right

Cover Page: Masterfile.
Title Page: Masterfile.
Content Page: Photoeuphoria/Dreamstime.

4: Sarsmis/Shutterstock; 7: Blamb/Shutterstock; 9: Temet/Istockphoto; 10: Sebastian Kaulitzki/Shutterstock; 11: Sebastian Kaulitzki/Shutterstock, Jubal Harshaw/Shutterstock; 12: CNRI/Science Photo Library; 13: Janice Carr/Centers for Disease Control and Prevention; 14: Steinhagen Artur/Shutterstock; 15: Sebastian Kaulitzki/Shutterstock; 16: Andrea Danti/Shutterstock; 17: Anatomical Design/Shutterstock, Cheryl Ann Quigley/Shutterstock, Catalin Petolea/Shutterstock, Sebastian Kaulitzki/Shutterstock; 18l: Sebastian Kaulitzki/Shutterstock; 18r: Sebastian Kaulitzki/Shutterstock; 19: Dennis Kunkel/Photolibrary; 21: Sebastian Kaulitzki/Shutterstock; 24: Gleb Semenjuk/Shutterstock, Tan4ikk/Shutterstock; 27: Art Vector/Shutterstock; 28: Art Vector/Shutterstock, Morgan Lane Photography/Shutterstock; 29: Tamara Bauer/Istockphoto; 30-31: Eric Erbe.

Q2AMedia Art Bank: Title Page, 5, 6, 8, 9, 10, 12, 20, 23.

12 11 10 9 8 7 6   14 15 16/0

Printed in the U.S.A.   40
First printing, January 2011

Table of Contents

Chapter 1

A Special Food Factory

Most food factories prepare foods by putting together the ingredients, baking or cooking them, and packaging them for you to eat! While these foods may look, smell, and taste good, your body cannot use them as they are.

Your body needs to be a *reverse* food factory. It has to break food into smaller particles that can be absorbed into your blood. The process through which your body breaks down food is known as **digestion**. The organs involved in digestion belong to the digestive system.

SMART WORD

digestion the process through which the body breaks down food into particles that can be used

4

The food you eat goes on quite a trip through your body. There are loops and bends, liquids that gush, and walls that squeeze. It sounds like a crazy ride at an amusement park — but it's not! It's the parts of your digestive system.

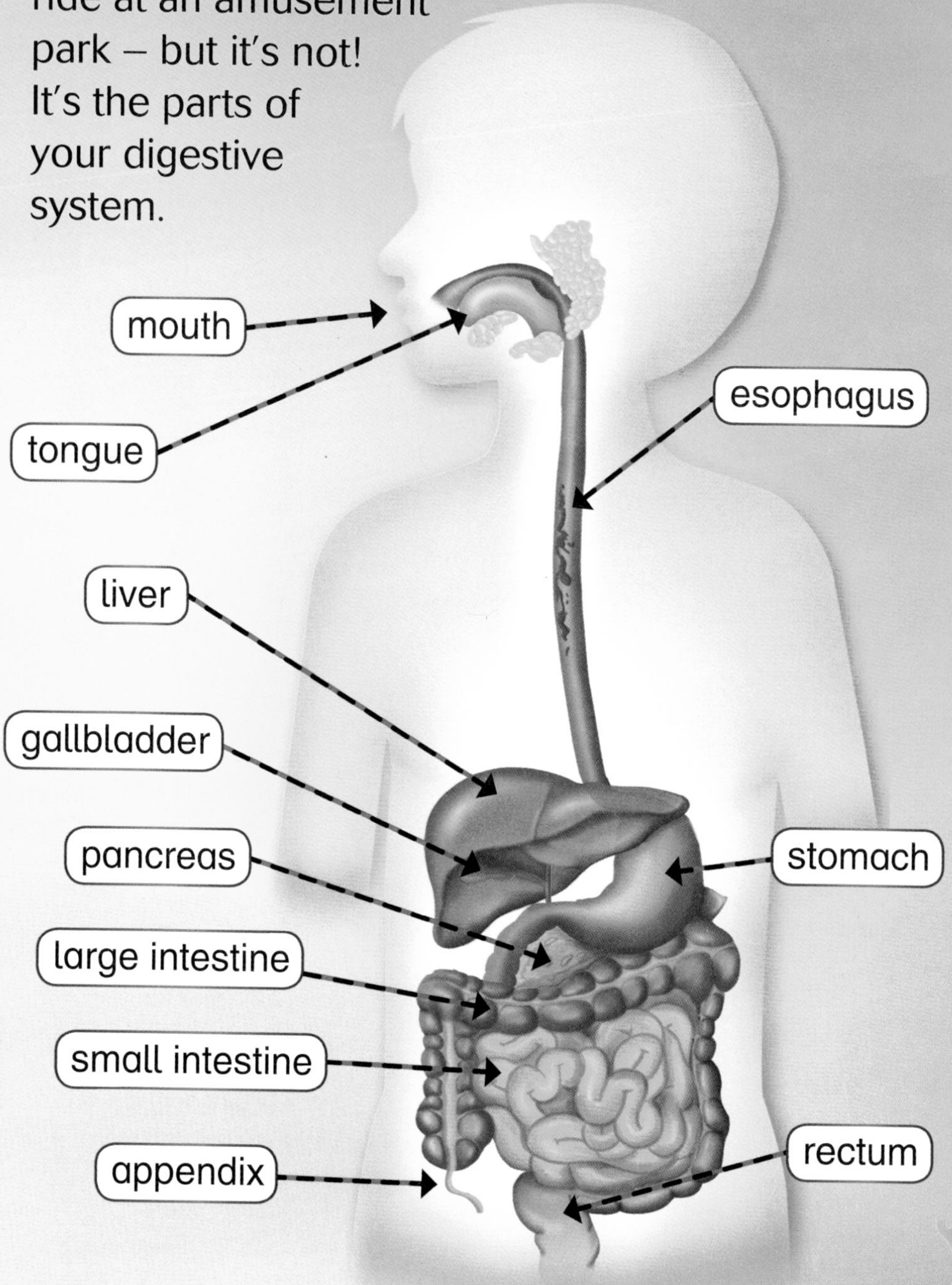

mouth

tongue

esophagus

liver

gallbladder

pancreas

stomach

large intestine

small intestine

appendix

rectum

Over the Lips and Into the Mouth

Believe it or not, the process of digestion begins before you even take your first bite of food! The smell of something delicious makes your mouth produce saliva. This watery liquid wets the food, making it easier to swallow. Saliva also contains **enzymes**, which are special proteins that help break down food.

Once you bite into food, the adventure really begins! Let's follow a turkey sandwich through the digestive system.

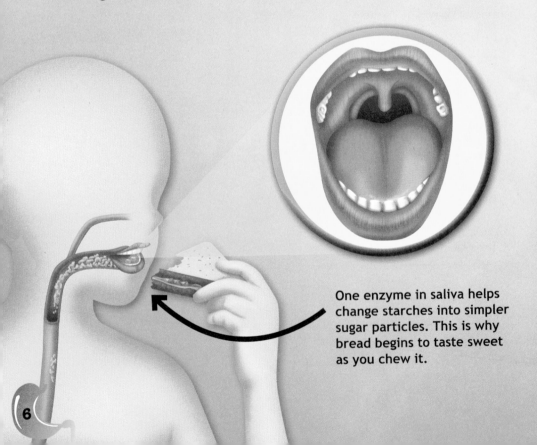

One enzyme in saliva helps change starches into simpler sugar particles. This is why bread begins to taste sweet as you chew it.

After a bit of chewing, you're ready to swallow. Your tongue pushes a bit of the food toward the back of your throat.

The throat has one passageway for air (windpipe) and one for food. A tissue called the epiglottis covers the opening to your windpipe. This keeps food from heading to your lungs.

tongue

epiglottis

People sometimes say that the "food went down the wrong pipe." This happens when food or liquids accidentally get into the windpipe. Coughing usually fixes the problem.

windpipe

enzymes   special proteins that help break down food during digestion

Down the Right Pipe

That bite of sandwich really needs to go into the **esophagus**. This is a tube that leads to the stomach. The esophagus is lined with muscles that squeeze in a wavelike pattern that moves the food along. These muscles would force food into the stomach even if you were hanging upside down. (But don't try it!)

The esophagus squeezes food down toward the stomach. A thick muscle stops the food from going back up.

Look Out Stomach, Here I Come

The **stomach** is a large, muscular sac. Once your sandwich gets there, the stomach acts like a mixer. It adds gastric juices that further break down your sandwich.

Muscles that line the stomach squeeze to churn the substances together. The mixture becomes known as chyme.

Food stays in the stomach between 2 and 4 hours. Then a valve opens to let it out, into the small intestine.

SMART WORDS

esophagus   the muscular tube through which food passes from the mouth to the stomach

stomach   the muscular, pouchlike organ that mixes food and chemicals together

Getting the Most Out of It

The **small intestine** continues the process of breaking down your turkey sandwich. To do this, it gets a little help from some other organs.

The **pancreas** produces enzymes that break down the remaining parts of your turkey sandwich even further. The **liver** produces bile. This liquid breaks up fat, such as the mayonnaise from your sandwich. Bile is stored in the gallbladder until it is needed.

You might be surprised to find out that the small intestine is only about 2 inches (5 centimeters) around.

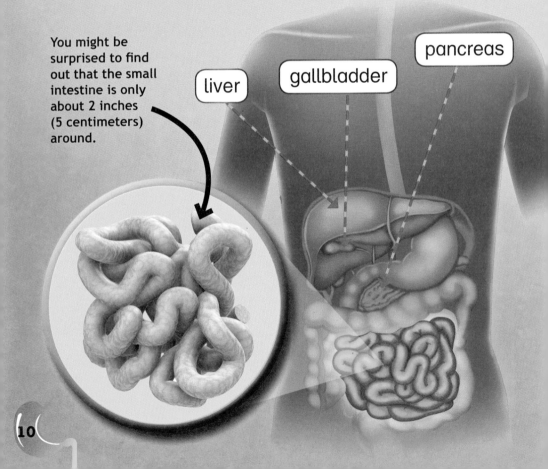

liver

gallbladder

pancreas

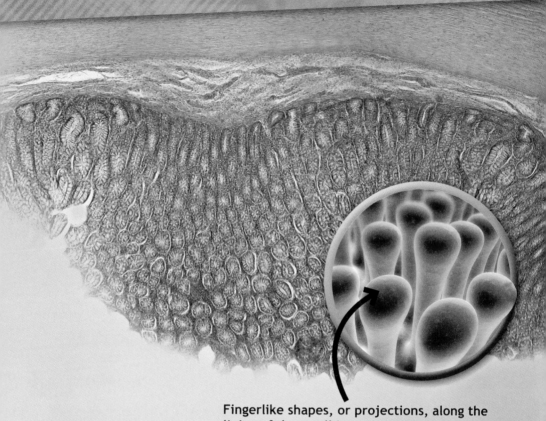

Fingerlike shapes, or projections, along the lining of the small intestine make it easier for food particles to pass into the blood.

Once the food is broken down into small particles, it can pass into the blood through the deeply folded walls of the small intestine. The blood then carries the nutrients to where they are needed in the body.

SMART WORDS

small intestine   the organ from which digested food particles are passed into the blood

pancreas   the organ that releases enzymes into the small intestine where they break down remaining food particles even further

liver   the organ that produces bile

Finishing the Job

All that is left in the small intestine now are materials that could not be digested, along with some water. These substances pass into the **large intestine**.

The large intestine is a tube-like organ through which undigested food can pass. The large intestine removes some water and minerals that remain in the material coming from the small intestine.

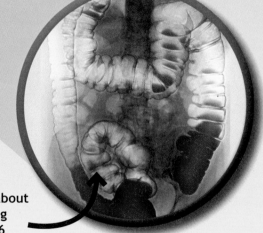

The large intestine is about 5 feet (1.5 meters) long and about 3 inches (7.6 centimeters) wide.

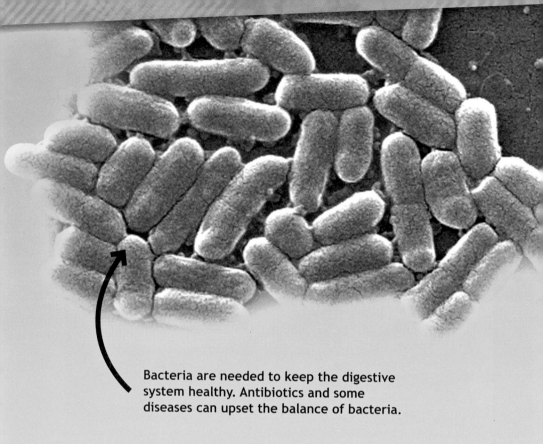

Bacteria are needed to keep the digestive system healthy. Antibiotics and some diseases can upset the balance of bacteria.

The large intestine is filled with helpful bacteria. These bacteria produce some materials that the body uses, such as vitamin K. Anything that is left by the time the material gets to the end of the large intestine is solid waste known as feces. Muscles in the lining of the large intestine contract to push the material out through the rectum.

SMART WORD

large intestine the tube-like organ through which undigested food passes

Use your SMART WORDS

Answer each question with a Smart Word.

> digestion enzyme liver stomach pancreas
>
> esophagus small intestine large intestine

1. I am the organ that produces bile to break down fat.

2. I am the organ that churns food and chemicals together like a mixer.

3. I am the organ that releases chemicals that break food into even smaller particles in the small intestine.

4. I am the process by which food is broken down into particles that can be absorbed by the body.

5. I am the tube that carries food from the mouth to the stomach.

6. I am a group of special proteins that help break down food during digestion.

7. I am the tube-like organ through which undigested food passes.

8. I am the organ in which digested food particles are passed into the blood.

Answers on page 32

Talk Like a Scientist

Can you describe this diagram of the digestive system? Use Smart Words to explain how food is broken down in each part of this system.

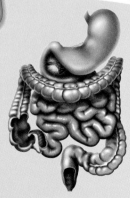

That's Amazing!

If stretched out, the small intestine is about 22 feet (6.7 meters) long and the large intestine is about 5 feet (1.5 meters) long.

Did You Know?

In a day, an adult digestive system processes about 3 gallons (11.4 liters) of food. Only about 4 ounces (120 milliliters) of that comes out as feces, or poop!

Incredible!

People produce about a half gallon (1.9 liters) of saliva every day! That's a lot of spit!

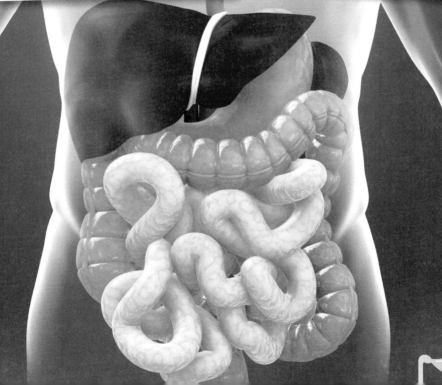

Chapter 2

Keeping the Balance

Think what would happen if trash piled up in your home. You would soon be unable to live there. The same is true in your body. Waste materials must be removed from your body to keep it healthy. The process of removing waste is known as **excretion**.The organs involved belong to the excretory system.

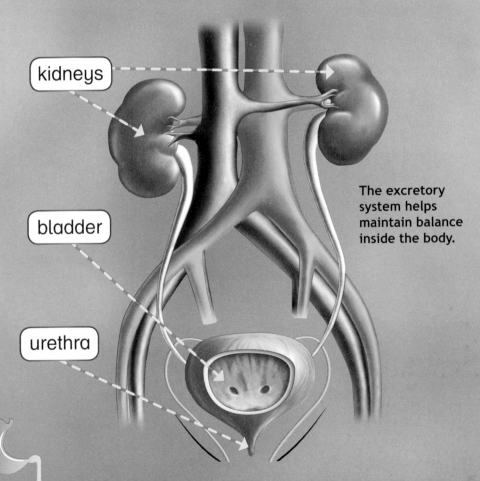

kidneys

bladder

urethra

The excretory system helps maintain balance inside the body.

The job of the excretory system is not just to clean up — it must also keep your body in balance! To keep you healthy, it must keep all of the internal conditions such as temperature, water content, and substances in your blood at proper levels. The process of keeping all these things in balance is called **homeostasis**.

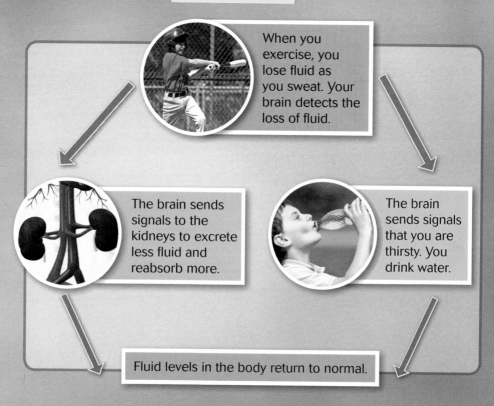

When you exercise, you lose fluid as you sweat. Your brain detects the loss of fluid.

The brain sends signals to the kidneys to excrete less fluid and reabsorb more.

The brain sends signals that you are thirsty. You drink water.

Fluid levels in the body return to normal.

SMART WORDS

excretion   the process of removing waste from the body

homeostasis   the process of keeping internal conditions of the body constant

Body Filters

Let's start looking at some organs of your excretory system and find out what they do.

A filter is something that blocks some materials from flowing through it. You may know about air filters or coffee filters. The **kidneys** are organs that filter waste out of the blood. Each of your kidneys contains millions of tiny structures called **nephrons**.

As blood passes through a nephron, substances such as water, salts, and food particles are filtered out.

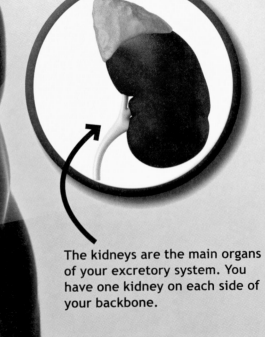

The kidneys are the main organs of your excretory system. You have one kidney on each side of your backbone.

Some of the filtered water and food is reabsorbed and returns to the body. The liquid that is left, called **urine**, passes into the **bladder**. It is stored there until it is removed from the body.

Nephrons in the kidneys can filter all of the blood in your body every 45 to 50 minutes!

SMART WORDS

kidney   one of two organs that filter waste from the blood and remove them as urine

nephron   a structure in a kidney that acts as a filter

urine   liquid produced by the kidneys to remove waste from the blood

bladder   organ that stores urine

Skin

You might not think your skin does much aside from covering your body. It actually does quite a lot! One important role is to excrete extra salts and water from the body. They are removed in **perspiration**, or sweat. Sweating not only cools you down, it also helps clean waste from your body.

Lungs

Take a deep breath in and out. Your lungs just brought oxygen into your body, but they also did something else. They removed waste. Every time you breathe out, you remove carbon dioxide and water vapor that formed during processes in your body.

One job of the lungs is to get oxygen into the body. Equally important is their job of getting rid of waste as you exhale, or breathe out.

Liver

In addition to the job the liver has in digestion, it also plays an important role in excretion. Sometimes your body produces harmful materials as part of its normal activities, such as breaking down certain foods. The liver changes them into other substances that can be safely filtered out by the kidneys.

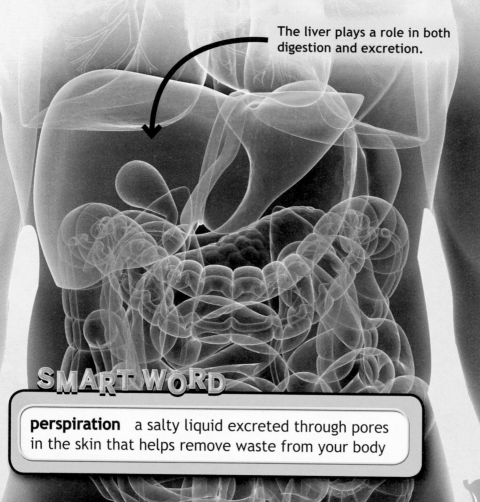

The liver plays a role in both digestion and excretion.

SMART WORD

perspiration   a salty liquid excreted through pores in the skin that helps remove waste from your body

Use your SMART WORDS

Match each description with the correct Smart Word.

excretion	bladder	urine	nephron
perspiration	homeostasis	kidney	

1. the organ that acts like a filter for the blood

2. the process of removing waste from the body

3. the process of keeping internal conditions of the body constant

4. the organ that stores urine until it is removed from the body

5. one of the many tiny filters inside each kidney

6. a salty fluid excreted through pores in the skin that helps remove wastes from your body

7. the liquid produced by the kidneys to remove waste from the blood

Answers on page 32

Talk Like a Scientist

Describe at least two ways waste materials can exit your body. Use your Smart Words.

The human body is 70 percent water.

Did You Know?

The human body is up to 70 percent water. A person can live about a month without food, but only one week without water.

That's Amazing!

Humans have about 10 pints (6 liters) of blood. The kidneys filter that blood about 400 times a day!

Looking for Clues

You can learn about water in your body by looking at the color of your urine. Urine that is dark yellow usually means you need to take in more water. Urine that is light yellow shows that the body has enough water and can get rid of some.

Chapter 3

The End Products

Wow! After making that trip through the digestive and excretory systems, that bite of turkey sandwich looks nothing like what it started as. What's the point of all that activity?

The answer is **energy**. Energy is basically the ability to do work or cause change. You need energy for everything you do – to walk down a hall, pick up your books, or read these words.

Your body cannot produce energy. Instead, you get the energy you need from the foods you eat. Food is like fuel for the body.

SMART WORDS

energy   the ability to do work or cause change

calorie   the unit used to measure the amount of energy in food

nutrient   substance that provides your body with energy and the other materials it needs

Nutrition Facts

Serving Size: 1 cup (228g)
Servings per Container: 2

① Start here ➡

② Check calories

Amount per Serving

Calories: 250   Calories from Fat: 110

	% Daily Value
Total Fat 12g	18%
Saturated Fat 3g	15%
Trans Fat 3g	
Cholesterol 30mg	10%
Sodium 470mg	20%
Total Carbohydrate 31g	10%
Dietary Fiber 0g	0%
Sugars 5g	
Protein 5g	
Vitamin A	4%
Vitamin C	2%
Calcium	20%
Iron	4%

③ **Limit these nutrients**

④ **Get enough of these nutrients**

This nutrition label shows the number of calories in one serving, as well as the nutrients in this food.

Different foods have different amounts of energy. The energy in food is measured in units called **calories**. The number of calories a person needs each day depends on his or her body size, level of activity, and age. A 7–10 year old needs an average of 2,000 calories a day, plus lots of physical activity.

You may have heard some foods described as having "empty calories." These are foods that are high in calories but low in nutrients. **Nutrients** are the substances that provide your body with the energy and other materials it needs.

Nutrients are so important that the United States Department of Agriculture (USDA) has established a guide for healthful eating. It is called the **Food Guide Pyramid**. Although there is no one diet that is right for every person, the pyramid acts as a guide for food and physical activity choices.

You can see that the pyramid is divided into sections. The size of the section tells you how much of your diet should be made up of each kind of food. For example, the grain section is the largest. A good part of your diet should consist of grains. The fats section is the smallest. Fats should make up the smallest part of your diet.

The stairs on the pyramid show that, in addition to eating the right foods, you need to keep your body moving by doing **exercise**.

SMART WORDS

Food Guide Pyramid   a guide developed for making healthful food and activity choices

exercise   physical activity that helps keep your body healthy

| grains 6–11 servings | vegetable 3–5 servings | fruit 2–4 servings | fats | milk 2–3 servings | meat & beans 2–3 servings |

Source: U.S. Department of Agriculture/U.S. Department of Health and Human Services

Make at least ½ of your grains "whole."

Eat a variety of fruits and vegetables.

Include calcium-rich foods such as low-fat milk.

Eat lean meats and ask for them broiled or grilled, instead of fried.

Make sure you drink plenty of water each day.

Balance your diet with plenty of exercise!

Match each description with the correct Smart Word.

> energy   calorie   nutrient
>
> exercise   Food Guide Pyramid

1. What is the ability to do work or cause change?

2. What is the unit used to measure the energy in food?

3. What is the guide for making healthful food choices?

4. What is a substance that provides your body with energy and the other materials it needs?

5. What is physical activity that helps keep your body healthy?

Answers on page 32

Talk Like a Scientist

Use the Food Guide Pyramid to plan three meals for a day. Use Smart Words to explain your food choices.

| grains 6–11 servings | vegetable 3–5 servings | fruit 2–4 servings | fats | milk 2–3 servings | meat & beans 2–3 servings |

SMART FACTS

Did You Know?

Roughly 40 percent of the total calories consumed by children ages 2–18 are in the form of empty calories from soda, desserts, pizza, and whole milk.

Think About It!

Over the past four decades, the rates of obesity have increased nearly five times among children ages 6–11. An estimated 25 million children are currently overweight or obese.

Good to Know!

The Department of Agriculture also has an Activity Pyramid. It suggests daily activities such as walking or bicycling instead of riding in a car or on a bus. Activities such as swimming, dancing, or sports should happen 3–5 times a week.

Glossary

bladder   organ that stores urine

calorie   the unit used to measure the amount of energy in food

digestion   the process through which the body breaks down food into particles that can be used

energy   the ability to do work or cause change

enzymes   special proteins that help break down food during digestion

esophagus   the muscular tube through which food passes from the mouth to the stomach

excretion   the process of removing waste from the body

exercise   physical activity that helps keep your body healthy

Food Guide Pyramid   a guide developed for making healthful food and activity choices

homeostasis   the process of keeping internal conditions of the body constant

kidney   one of two organs that filter waste from the blood and remove them as urine

large intestine   the tube-like organ through which undigested food passes

liver   the organ that produces bile

nephron   a structure in a kidney that acts as a filter

nutrient   substance that provides your body with energy and the other materials it needs

pancreas   the organ that releases enzymes into the small intestine where they break down remaining food particles even further

perspiration   a salty liquid excreted through pores in the skin that helps remove waste from your body

small intestine   the organ from which digested food particles are passed into the blood

stomach   the muscular, pouchlike organ that mixes food and chemicals together

urine   liquid produced by the kidneys to remove waste from the blood

Index

SMART WORDS Answer Key

Page 14
1. liver, 2. stomach, 3. pancreas, 4. digestion, 5. esophagus, 6. enzyme, 7. large intestine, 8. small intestine

Page 22
1. kidney, 2. excretion, 3. homeostasis, 4. bladder, 5. nephron, 6. perspiration, 7. urine

Page 28
1. energy, 2. calorie, 3. Food Guide Pyramid, 4. nutrient, 5. exercise